SPARKS OF LIFE

Chemical Elements that Make Life Possible

SULFUR

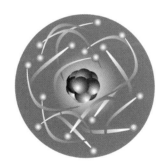

by

Jean F. Blashfield

RAINTREE
STECK-VAUGHN
PUBLISHERS

A Harcourt Company

Austin · New York
www.steck-vaughn.com

Special thanks to our technical consultant,
Philip T. Johns, Ph.D.
University of Wisconsin—Whitewater, Wisconsin

NL

Development: Books Two, Delavan, Wisconsin
        Graphics: Krueger Graphics, Janesville, Wisconsin
        Interior Design: Peg Esposito
        Photo Research: Margie Benson
        Indexing: Winston E. Black

Raintree Steck-Vaughn Publisher's Staff:
        Publishing Director: Walter Kossmann   Project Editor: Sean Dolan
        Design Manager: Max Brinkmann   Electronic Production: Scott Melcer

Library of Congress Cataloging-in-Publication Data:
Blashfield, Jean F.
  Sulfur / by Jean F. Blashfield.
    p. cm. — (Sparks of life)
  Includes bibliographical references and index.
  ISBN 0-7398-3452-5
    1. Sulphur--Juvenile literature. 2. Sulphur--Physiological aspects--Juvenile
literature. [1. Sulphur.] I. Title.

  QD181.S1. B53 2001
  545'.723--dc21   00-045688

Printed and bound in the United States
1 2 3 4 5 6 7 8 9 LB 05 04 03 02 01

PHOTO CREDITS: B.I.F.C. cover; ©Jean Black 48; ©Jules Bucher/Photo Researchers 37;
©David Cavagnaro/Peter Arnold Inc. 9; ©Doug Cheeseman/Peter Arnold Inc. 18;
©Vaughan Fleming/Science Photo Library/Photo Researchers 56; ©1998 Ross Fried/
Visuals Unlimited 12; ©Terry Gleason/Visuals Unlimited 29; The Goodyear Tire &
Rubber Company 55; ©Holt Studios/Nigel Cattlin/Photo Researchers 45; R. Margulies/
Custom Medical Stock Photo 50; ©Joe McDonald/Visuals Unlimited 35; National
Renewable Energy Laboratory 30, 57; ©Lawrence Naylor/Photo Researchers 32;
©NCA/Visuals Unlimited 36; NOAA/PMEL /Vents (NOAA Pacific Marine Environmental
Lab, NOAA Hydrothermal Vents Program 24; NU Institute of Agriculture & Natural
Resources 43; ©Alfred Pasieka/Science Photo Library 22; ©Harry J. Przekop, Stock
Shop/Medichrome cover; ©Michael V. Propp/Peter Arnold Inc cover; ©D. Van
Ravenswaay/Science Photo Library/Photo Researchers 40; ©Mark A. Schneider/Visuals
Unlimited cover; ©Science Vu/Visuals Unlimited 38; ©Paul A. Souders/CORBIS 25;
©Inga Spence/Visuals Unlimited 26, 47; ©Jim Steinberg/Photo Researchers 31; ©L.S.
Stepanowicz/Visuals Unlimited 17; ©David Taylor/Science Photo Library/Photo Re-
searchers 16; ©R. Treptow/Visuals Unlimited 15; ©Tom Ulrich/Visuals Unlimited cover,
41; ©Dave Weintraub/Photo Researchers 19; ©Jim Yokajty/The Image Finders 10.

18 95

CONTENTS

S

Periodic Table of the Elements

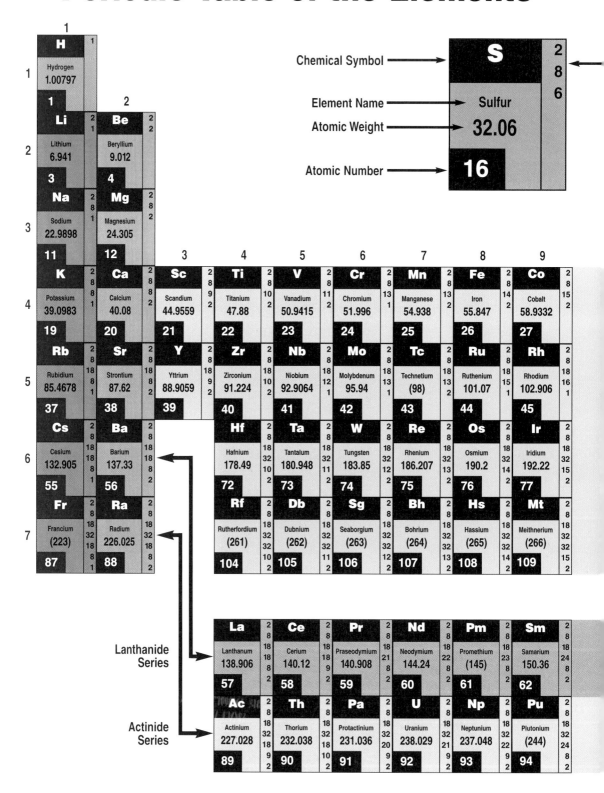

Chemical Symbol → **S** — 2 8 6

Element Name → Sulfur

Atomic Weight → 32.06

Atomic Number → 16

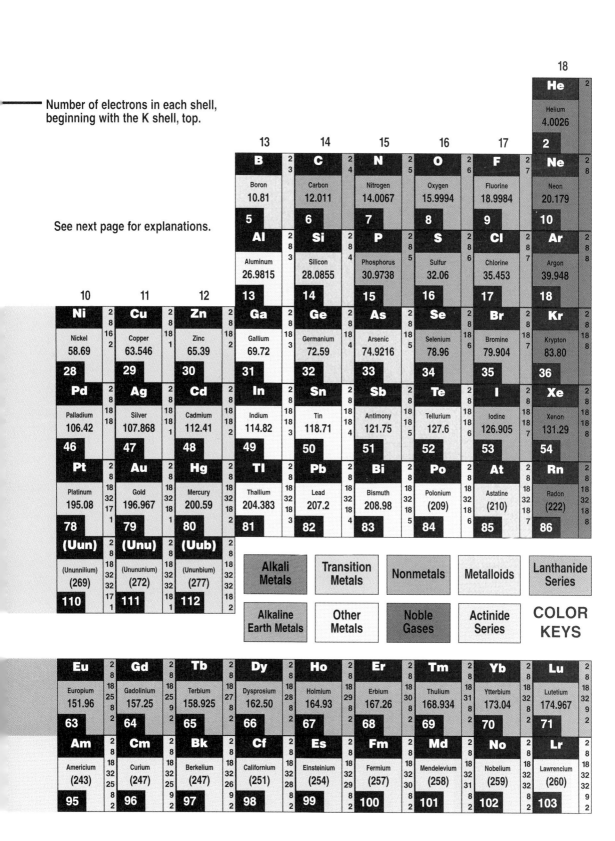

Number of electrons in each shell, beginning with the K shell, top.

See next page for explanations.

18

	13	14	15	16	17	18

He — 2
Helium
4.0026
2

13	14	15	16	17	

B — 2, 3
Boron
10.81
5

C — 2, 4
Carbon
12.011
6

N — 2, 5
Nitrogen
14.0067
7

O — 2, 6
Oxygen
15.9994
8

F — 2, 7
Fluorine
18.9984
9

Ne — 2, 8
Neon
20.179
10

Al — 2, 8, 3
Aluminum
26.9815
13

Si — 2, 8, 4
Silicon
28.0855
14

P — 2, 8, 5
Phosphorus
30.9738
15

S — 2, 8, 6
Sulfur
32.06
16

Cl — 2, 8, 7
Chlorine
35.453
17

Ar — 2, 8, 8
Argon
39.948
18

10	11	12	13	14	15	16	17	18

Ni — 2, 8, 16, 2
Nickel
58.69
28

Cu — 2, 8, 18, 1
Copper
63.546
29

Zn — 2, 8, 18, 2
Zinc
65.39
30

Ga — 2, 8, 18, 3
Gallium
69.72
31

Ge — 2, 8, 18, 4
Germanium
72.59
32

As — 2, 8, 18, 5
Arsenic
74.9216
33

Se — 2, 8, 18, 6
Selenium
78.96
34

Br — 2, 8, 18, 7
Bromine
79.904
35

Kr — 2, 8, 18, 8
Krypton
83.80
36

Pd — 2, 8, 18, 18
Palladium
106.42
46

Ag — 2, 8, 18, 18, 1
Silver
107.868
47

Cd — 2, 8, 18, 18, 2
Cadmium
112.41
48

In — 2, 8, 18, 18, 3
Indium
114.82
49

Sn — 2, 8, 18, 18, 4
Tin
118.71
50

Sb — 2, 8, 18, 18, 5
Antimony
121.75
51

Te — 2, 8, 18, 18, 6
Tellurium
127.6
52

I — 2, 8, 18, 18, 7
Iodine
126.905
53

Xe — 2, 8, 18, 18, 8
Xenon
131.29
54

Pt — 2, 8, 18, 32, 17, 1
Platinum
195.08
78

Au — 2, 8, 18, 32, 18, 1
Gold
196.967
79

Hg — 2, 8, 18, 32, 18, 2
Mercury
200.59
80

Tl — 2, 8, 18, 32, 18, 3
Thallium
204.383
81

Pb — 2, 8, 18, 32, 18, 4
Lead
207.2
82

Bi — 2, 8, 18, 32, 18, 5
Bismuth
208.98
83

Po — 2, 8, 18, 32, 18, 6
Polonium
(209)
84

At — 2, 8, 18, 32, 18, 7
Astatine
(210)
85

Rn — 2, 8, 18, 32, 18, 8
Radon
(222)
86

(Uun) — 2, 8, 18, 32, 17, 1
(Ununnilium)
(269)
110

(Unu) — 2, 8, 18, 32, 18, 1
(Unununium)
(272)
111

(Uub) — 2, 8, 18, 32, 18, 2
(Ununbium)
(277)
112

Alkali Metals	Transition Metals	Nonmetals	Metalloids	Lanthanide Series
Alkaline Earth Metals	Other Metals	Noble Gases	Actinide Series	**COLOR KEYS**

Eu — 2, 8, 18, 25, 8, 2
Europium
151.96
63

Gd — 2, 8, 18, 25, 9, 2
Gadolinium
157.25
64

Tb — 2, 8, 18, 27, 8, 2
Terbium
158.925
65

Dy — 2, 8, 18, 28, 8, 2
Dysprosium
162.50
66

Ho — 2, 8, 18, 29, 8, 2
Holmium
164.93
67

Er — 2, 8, 18, 30, 8, 2
Erbium
167.26
68

Tm — 2, 8, 18, 31, 8, 2
Thulium
168.934
69

Yb — 2, 8, 18, 32, 8, 2
Ytterbium
173.04
70

Lu — 2, 8, 18, 32, 9, 2
Lutetium
174.967
71

Am — 2, 8, 18, 32, 25, 8, 2
Americium
(243)
95

Cm — 2, 8, 18, 32, 25, 9, 2
Curium
(247)
96

Bk — 2, 8, 18, 32, 26, 9, 2
Berkelium
(247)
97

Cf — 2, 8, 18, 32, 28, 8, 2
Californium
(251)
98

Es — 2, 8, 18, 32, 29, 8, 2
Einsteinium
(254)
99

Fm — 2, 8, 18, 32, 30, 8, 2
Fermium
(257)
100

Md — 2, 8, 18, 32, 31, 8, 2
Mendelevium
(258)
101

No — 2, 8, 18, 32, 32, 8, 2
Nobelium
(259)
102

Lr — 2, 8, 18, 32, 32, 9, 2
Lawrencium
(260)
103

A Guide to the Periodic Table

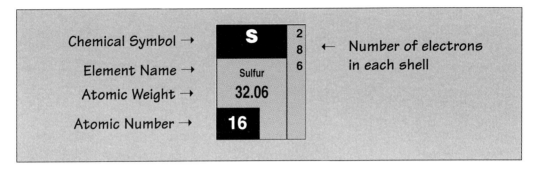

Symbol = an abbreviation of an element name, agreed on by members of the International Union of Pure and Applied Chemistry. The idea to use symbols was started by a Swedish chemist, Jöns Jakob Berzelius, about 1814. Note that the elements with numbers 110, 111, and 112, which were "discovered" in 1996, have not yet been given official names.

Atomic number = the number of protons (particles with a positive electrical charge) in the nucleus of an atom of an element; also equal to the number of electrons (particles with a negative electrical charge) found in the shells, or rings, of an atom that does not have an electrical charge.

Atomic weight = the weight of an element compared to carbon. When the Periodic Table was first developed, hydrogen was used as the standard. It was given an atomic weight of 1, but that created some difficulties, and in 1962, the standard was changed to carbon-12, which is the most common form of the element carbon, with an atomic weight of 12.

The Periodic Table on pages 4 and 5 shows the atomic weight of carbon as 12.011 because an atomic weight is an average of the weights, or masses, of all the different naturally occurring forms of an atom. Each form, called an isotope, has a different number of neutrons (uncharged particles) in the nucleus. Most elements have several isotopes, but chemists assume that any two samples of an element are made up of the same mixture of isotopes and thus have the same mass, or weight.

Electron shells = regions surrounding the nucleus of an atom in which the electrons move. Historically, electron shells have been described as orbits similar to a planet's orbit. But actually they are whole areas of a specific energy level, in which certain electrons vibrate and move around. The shell closest to the nucleus, the K shell, can contain only 2 electrons. The K shell has the lowest energy level, and it is very hard to break its electrons away. The second shell, L, can contain only 8 electrons. Others may contain up to 32 electrons. The outer shell, in which chemical reactions occur, is called the valence shell.

Periods = horizontal rows of elements in the Periodic Table. A period contains all the elements with the same number of orbital shells of electrons. Note that the actinide and lanthanide (or rare earth) elements shown in rows below the main table really belong within the table, but it is not regarded as practical to print such a wide table as would be required.

Groups = vertical columns of elements in the Periodic Table; also called families. A group contains all elements that naturally have the same number of electrons in the outermost shell or orbital of the atom. Elements in a group tend to behave in similar ways.

Group 1 = alkali metals: very reactive and so never found in nature in their pure form. Bright, soft metals, they have one valence electron and, like all metals, conduct both electricity and heat.

Group 2 = alkaline earth metals: also very reactive and thus don't occur pure in nature. Harder and denser than alkali metals, they have two valence electrons that easily combine with other chemicals.

Groups 3–12 = transition metals: the great mass of metals, with a variable number of electrons; can exist in pure form.

Groups 13–17 = transition metals, metalloids, and nonmetals. Metalloids possess some characteristics of metals and some of nonmetals. Unlike metals and metalloids, nonmetals do not conduct electricity.

Group 18 = noble, or rare, gases: in general, these nonmetallic gaseous elements do not react with other elements because their valence shells are full.

MAGIC AND BRIMSTONE

Some early humans lived near volcanoes. These people knew that a substance that looked like burning rock and had a bad odor sometimes escaped from these volcanoes. These early people also saw great deposits of a vivid yellow material build up along streams near volcanoes. Some of these early people used the yellow substance as paint to draw pictures on the walls of caves.

Perhaps these early people began to consider this yellow material as magical, especially when it caught on fire. Eventually, this yellow burning rock was called brimstone, and the terrible underworld where brimstone came from was thought to be Hell. The Book of Genesis in the Bible says the ancient cities of Sodom and Gomorrah were destroyed by fire and brimstone.

Yellow sulfur accumulates around the openings of steam vents near the top of a volcano.

Gradually, the yellow substance, which we know today as sulfur, was found to be an ingredient in many rocky ores from which people obtained metals. The ancient Greeks thought sulfur was the "soul" of these metals. Thus, the Greek word for sulfur, *theion,* was also the Greek word for "god" or "divinity." The Greeks also thought that mercury (Hg, element #80 on the Periodic Table), which was also common in metal ores, was the "intelligence" of metals.

Today we know that sulfur is a chemical element—one of the basic building blocks of all matter. The atomic symbol for sulfur is S and it is element #16 on the Periodic Table of the Elements (see pages 4 and 5).

In ancient Rome, sulfur was used to bleach wool. Bleaching makes a colored material white. Fabric made from sheeps' wool was naturally yellow-gray. The Romans would spread the fabric above a pot of burning sulfur. As the sulfur burned, it combined with the element oxygen (O, element #8) in air to form a new substance called sulfur dioxide, which is written SO_2. The formula SO_2 means sulfur dioxide is composed of one part sulfur (S) and two parts oxygen (O_2). The fumes of sulfur dioxide removed color from the sheeps' wool, bleaching it white. Sulfur dioxide is still used today to bleach fabrics.

A street in ancient Pompeii, which was once covered by lava from Mount Vesuvius, visible in the background. The people of Pompeii were killed by poisonous sulfur dioxide before being buried.

Sulfur dioxide can kill people, as the ancient Romans learned when the residents of the ancient towns of Pompeii and Herculaneum were killed by breathing in sulfur dioxide. Pompeii and Herculaneum were resort towns in Italy where the rich people of Rome went for vacations. The two towns were completely covered in lava after the sudden eruption of the volcano Mount Vesuvius in the year A.D. 79. But before the thousands of residents were buried by lava, they were probably already dead from breathing the sulfur dioxide gas given off by the volcano.

Getting to Know the Element

When ancient people used sulfur, they did not know they were dealing with a chemical element, one of the basic substances of the universe. They probably thought of sulfur as just that volcanic stuff that was sometimes yellow, sometimes smelly, sometimes suffocating, and sometimes useful. Because it burned so easily, the word *sulfur* came to mean any chemical that easily burned. Therefore, when the word appears in old texts, it is not certain whether the writers were referring to the element or to another material that was being studied.

The ancient Chinese used sulfur as an ingredient in an explosive powder they used to make colorful fireworks. Later, this explosive was called black powder because one of the prime ingredients in addition to sulfur was charcoal, which is a combination of carbon (C, element #6) and other materials. Charcoal is black.

Making fireworks and blasting holes in the earth for mining were probably the only uses black powder had for centuries. Then, sometime after 1100, people began using black powder as gunpowder. Either on their own or having learned about Chinese experiments, military experts in Europe also developed weapons that used gunpowder. They could shoot an iron ball out of a heavy chamber or gun barrel and send the ball long distances. Cannons were first used in European warfare in 1346 at the Battle of Crècy during the Hundred Years' War.

Gunpowder quickly became so important that when the Spaniards conquered Mexico in the early 1500s, Hernando Cortès, a Spanish military leader, found a dangerous way to obtain the sulfur his army needed to make gunpowder. He sent daring soldiers on ropes down into a Mexican volcano to obtain sacks of sulfur from the deposits inside the volcano's cone.

More than 200 years later, in 1777, French chemist Antoine Lavoisier, known for his study of many elements, declared sulfur to be a chemical element. He showed that it could not be broken down into smaller component parts and still maintain the characteristics of sulfur.

Many people spell the name of the element *sulphur*. But historical references to the element show it was spelled *sulfur* just as often as it was spelled *sulphur*. Recently, the International Union of Pure and Applied Chemistry stated that the proper spelling of the element should be with an *f*.

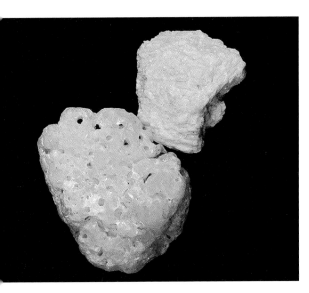

Pure sulfur from an ancient volcano in Utah. The larger chunk has changed slightly just from being held in a warm hand.

The Reactive Atom

Sulfur's atomic number is 16. This means that sulfur has 16 protons, or positively charged particles. It also has 16 neutrons, or particles with no electrical charge, in its nucleus. In addition, 16 electrons, or negatively charged particles, move in regions called orbits, or shells, around its nucleus. These electrons are located in three different shells. This fact places sulfur in the horizontal row of the Periodic Table of the Elements called Period 3.

There are two electrons in sulfur's first, or inner, shell; eight in its second, or middle shell; and six in its outer, or valence, shell, which is where chemical reactions take place. This fact places it in Group 16 (also called Group VIA) on the Periodic Table. First in this column is oxygen. Other elements in Group 16 are selenium (Se, element #34), tellurium (Te, #52), and polonium (Po, #84).

Theoretically, an element can hold up to eighteen electrons in that third shell. But with only six electrons in its valence shell, we say that sulfur is "unstable," that is, it tends to react quickly with other elements, borrowing electrons from them. Elements are stable when they have eight or eighteen electrons in that outer shell. A stable element reacts less quickly with other elements or compounds.

All elements try to become stable by adding missing electrons to the valence shell or by giving up electrons if there are only a few in the valence shell. Therefore, sulfur has a natural tendency to react quickly with, or share, electrons from other elements. Sulfur needs only two more electrons, for a total of eight in its outer shell, to become stable. This means that it will quickly react with any atom that can share two electrons or with two atoms that can share one each.

For example, sulfur combines quickly with two atoms from hydrogen (H, element #1) to form a compound called hydrogen sulfide, H_2S. The formula for H_2S means that this gas is made up of two atoms of hydrogen (H_2) and one atom of sulfur (S). Hydrogen sulfide is the gas that gives off a terrible odor that people often refer to as the "smell of rotten eggs."

The Burning Stone

When pure sulfur is exposed to air, nothing happens. But if the temperature of the sulfur is raised only slightly, it will burn with a bright blue flame. The fact that sulfur burns easily gave it the name "brimstone," meaning "burning stone." The ability of sulfur to burn rapidly in air allowed ancient peoples to use sulfur as a type of match to light fires, long before the matches we know today were invented.

The reaction of sulfur and hydrogen achieve stability.

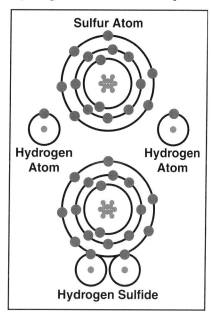

When sulfur is heated to its boiling point of 444.6°C (831°F), it first melts and then turns to a gas. If it is then cooled quickly, it changes directly to a solid instead of returning to a liquid form first. This is a process called sublimation. The result is a finely powdered form of sulfur called flowers of sulfur. Flowers of sulfur is often used as a fertilizer.

Sulfur's Isotopes

Sulfur atoms can occur in different forms called isotopes. The word *isotope* means "same place." All isotopes of an element occupy the "same place" on the Periodic Table.

The most common isotope of naturally occurring sulfur (about 95 percent) has 16 neutrons and 16 protons in its nucleus. The 16 neutrons and 16 protons of this isotope of sulfur give it an atomic weight, or mass, of 32, often written S-32.

However, different isotopes of sulfur have different numbers of neutrons in their nuclei. Some atoms of sulfur exist with 17 neutrons (S-33), 18 (S-34), and 19 (S-35), giving these isotopes atomic weights of 33, 34, and 35. When these four isotopes are averaged together in the amounts they occur, sulfur as a whole has an atomic weight of 32.06. Deposits of sulfur contain mostly S-32, but different places on Earth have sulfur containing different amounts of the other isotopes.

Sulfur also has other isotopes that were created artificially by scientists bombarding sulfur atoms with neutrons in huge machines. Unlike natural isotopes, these isotopes of sulfur are radioactive—they give off high-energy particles.

Crowns and Crystals

Pure, or elemental, sulfur does not exist as individual atoms, as many elements do. Instead, elemental sulfur consists of molecules made up of eight atoms. The eight atoms are connected in a closed zigzag chain.

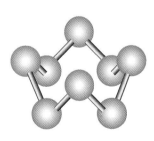

A sulfur molecule

Crystals are solids in which the atoms form regular repeating three-dimensional patterns. If you have ever looked closely at table salt, you have seen square crystals of sodium chloride, NaCl (sodium is Na, element #11; chlorine is Cl, #17). Crystals of elemental

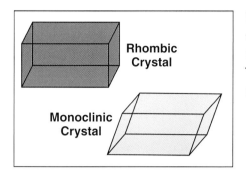

Rhombic Crystal

Monoclinic Crystal

Rhombic crystals of sulfur can easily be melted to a liquid. They cool into monoclinic crystals.

sulfur look different. In fact, they occur in two different types of crystals.

The most stable form of sulfur is rhombic sulfur. The crystals are shaped like the geometric figure called a rhombus, as if the crown-shaped rings of sulfur molecules were piled on top of one another. Rhombic sulfur is dark yellow, almost amber.

If rhombic crystals are heated almost to the melting point, as they might be inside a volcano, the sulfur changes shape, into what is called a monoclinic crystal. The molecules of sulfur still have eight atoms, but they have shifted so that the crystal is shaped like a rectangular box that has partly collapsed sideways. Monoclinic sulfur crystals, which look needle-like, are lighter yellow than rhombic sulfur.

Changing Temperatures, Changing Forms

Sulfur is an intriguing element for chemists. Rhombic and monoclinic sulfur are just two of the many different forms, or allotropes, that sulfur can take. Allotropes are different from isotopes. Isotopes involve differences in the nucleus, while allotropes are differences in the forms that the element itself can take. For example, two of carbon's allotropes are diamonds and coal—very different but both made of the same kind of atoms.

Sulfur has more allotropes than any other element—at least nineteen. Allotropes form when sulfur is melted quickly or

Sulfur heated to the boiling point turns dark red and plastic.

slowly. When the crystals first melt, they form an easily flowing fluid. But when the temperature is raised to between 160° and 195°C (320°–383°F), it becomes so thick and rubbery that it does not flow. The crown shapes of the sulfur crystals open and link up with other sulfur atoms, forming long, tangled chains of atoms. If the substance is heated to the boiling point, the chains break up and the fluid, once again flowing, becomes dark red, almost black, in color.

Most important is the fact that when the crown shapes of sulfur crystals open up, the atoms are free to share their valence electrons with other elements. Almost all elements can combine with sulfur, making it one of the most widely dispersed elements in nature.

Sulfur also exists in a non-crystalline allotrope called *amorphous* (meaning "formless") sulfur, sometimes called plastic sulfur. Molecules of this allotrope are made of eight atoms, just as sulfur crystals are, but each group of eight atoms forms a spiral in a long chain of linked spirals.

The Oxygen Scavenger

Sulfur has been called an oxygen scavenger because of the ease with which oxygen atoms in other compounds leave those compounds and attach themselves to the sulfur. It is as an oxygen scavenger that sulfur is used in many industrial processes.

Sulfur forms several compounds with oxygen. Collectively, these compounds are known as SO_x. When sulfur is burned in air, it combines with two atoms of oxygen, making sulfur dioxide, SO_2 (one atom of sulfur and two atoms of oxygen).

SO_2 is a gas at normal temperatures. However, below 10°C

(14°F), it becomes a liquid. Liquid SO_2 is stored in high-pressure cylinders and sold to the industries that use it. Sulfur dioxide, which has a bad odor, is a very harmful compound, especially if it gets into the air we breathe. When sulfur dioxide enters the atmosphere, it quickly changes to sulfur trioxide, SO_3. This is a corrosive gas that chokes anyone breathing it.

Other compounds of oxygen and sulfur form ions, which are molecules that have an electrical charge because they are missing an electron or two or have an extra electron or two. Sulfur ions all have negative charges because they have excess electrons. The atoms of these sulfur ions tend to remain together when they combine with other atoms.

Two important compounds of sulfur made with ions are sulfites and sulfates. A sulfite contains the SO_3^{2-} ion. The $^{2-}$ in the formula means that this ion has two extra electrons. In forming a molecule, a sulfite may give one electron to each of two atoms. For example, sodium sulfite is Na_2SO_3. It takes two atoms of sodium to use up the two extra electrons in the sulfite ion. Sodium sulfite is used as a bleach and a food preservative.

Sulfates are similar to sulfites, but they contain the SO_4^{2-} ion. The mineral calcium sulfate, $CaSO_4$ (calcium is Ca, element #20), is gypsum, used in making drywall and plaster casts.

Sulfur is a busy element, involved in many ways in Earth and its living creatures. Many scientists regard it second only to the element carbon in importance.

Gypsum is calcium sulfate, $CaSO_4$. It is found naturally in deposits left by ancient oceans. For centuries, it has been powdered and mixed with a little water to make plaster for walls.

VOLCANOES, VENTS, AND THAT YELLOW STUFF

Sulfur is in both the flaming molten lava and the invisible gases that spew out of a volcano.

Sulfur is the ninth most abundant element in the universe. There's about the same amount of it as there is of iron (Fe, element #9).

Most of the elements making up Earth still reside primarily in the hot, melted rock below the planet's crust, or outer surface. The crust "floats" on the molten rock on huge plates called tectonic plates. Some molten rock comes to the surface—sometimes oozing, sometimes exploding—at the weak spots in the crust where two tectonic plates meet each other.

Natural steam vents, or holes, located at thin spots in the crust often give off hot gases, usually foul-smelling hydrogen sulfide, H_2S. Yellowstone National Park has many such vents.

Mount St. Helens in Washington state was thought to be extinct until it erupted unexpectedly in 1980.

One of the places in the Pacific made famous during World War II is the island of Iwo Jima. The island was known to the Japanese as Sulfur Island because of its numerous vents. During February and March of 1945, one of the worst battles of the war took place on Iwo Jima between U.S. Marines and Japanese troops.

In other places, instead of forming vents, the gases build up underground until they erupt into an explosion of gases and hot lava. The lava builds up, forming a volcano. Many mountains in Earth's crust were originally active volcanoes. Sometimes, volcanoes that were thought to be extinct—incapable of erupting again—surprise people by erupting again. Mount St. Helens in Washington state did that in 1980, after being thought extinct for more than a hundred years.

The gases that explosively spew from volcanoes contain massive amounts of hydrogen sulfide. It reacts with the oxygen in the air to form water, which we see as steam, and sulfur:

$$2H_2S + O_2 \rightarrow 2H_2O + S$$

It is this sulfur that forms the huge yellow deposits found at various places on the planet.

Humans can safely inhale only very small amounts of

hydrogen sulfide. If there is much of it in the air, it can actually kill people by stopping their respiratory systems. People in laboratories who work with hydrogen sulfide must be extremely careful. Its presence in air can deaden the sense of smell, so that people are no longer aware that it is in the air around them. By the time they realize that something is wrong, it may already be too late.

Obtaining Elemental Sulfur

Sulfur became so important to industry in the 1800s that the readily available deposits on volcanoes were quickly used up. Important additional deposits were found on the Italian island of Sicily, where sulfur lay near the surface. Sicily was the main source of the element until the 1880s. Deposits were then found in the U.S. state of Louisiana near the Gulf of Mexico. These sulfur deposits existed deep underground in so-called salt domes, which are deposits of sodium chloride, NaCl. These deposits of sulfur were as much as 45.7 meters (150 feet) thick.

German-American chemist Herman Frasch, who worked with the early petroleum industry in Pennsylvania, became interested in sulfur because the first petroleum he worked with smelled bad due to the sulfur compounds in it. He devised ways to remove the sulfur to make the oil smell better, so that people would be more likely to buy it and use it in their homes.

Frasch built on what he knew about how petroleum is pumped out of the ground. Elemental sulfur does not dissolve in water, but water can be used to get it out of the ground. Extra-hot water is forced through a pipe into the thick layer of sulfur underground. The sulfur melts at 112.8°C (235°F), and the yellow melted liquid is forced by compressed air through another pipe up to the surface.

By 1902, Frasch had worked out all the problems, and the United States was obtaining cheap sulfur. It was important that

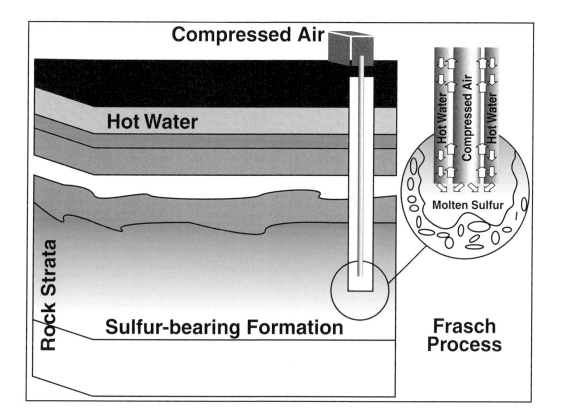

Compressed Air

Hot Water

Rock Strata

Sulfur-bearing Formation

Hot Water

Compressed Air

Hot Water

Molten Sulfur

Frasch Process

sulfur be cheap because sulfuric acid, H_2SO_4, was already becoming the most common chemical used in industry, especially for manufacturing fertilizers.

Moving Through the Earth

The sulfur that was present in Earth as it was formed is still the sulfur we have today. It has changed its form many times. This movement over time is often called the sulfur cycle because the element moves through Earth, its waters, and its atmosphere, often coming back to where it started.

The sulfur cycle begins in the molten rock under Earth's crust. Sulfur comes to the surface in volcanoes and in the water from hot springs. Both volcanic regions and the area around geysers and other hot springs often look yellow from sulfur.

Some sulfur blended into rocks, usually as sulfides, which are compounds that contain the sulfide ion S^{2-}. Pyrite, which is ferric, or iron, sulfide, FeS_2, is the most common sulfur mineral. Pyrite is found in sedimentary rocks, which are rocks that have formed out of the substances that settled to the bottom of bodies of water. Because sulfur makes the rocks look golden, pyrite is sometimes called "fool's gold" for the ease with which it fooled prospectors looking for gold.

Pyrite crystals make rock look as if it contains gold.

In addition to pyrite, many other sulfide minerals are sources of important metals. For example, galena or lead sulfide, PbS, provides a source of lead (Pb, element #82). Chalcopyrite or copper sulfide, CuS, provides a source of copper (Cu, #29). Cinnabar or mercury sulfide, HgS, is the main source of the silvery fluid element called mercury. Sulfur combines readily with all metallic elements except gold (Au, #79) and platinum (Pt, #78).

Sulfur gets into ocean water because sulfur-containing rocks under the sea and on land wear away, or weather. Sulfur grains and compounds are carried into rivers and then into the seas.

Living things also include sulfur. The element is essential to the normal functioning of both plants and animals. Millions of years ago, when much of the land was swamp, great beds of fallen plants were flooded and compressed. These gradually turned into the great deposits of coal, petroleum, and natural gas that fuel our world today. All these fossil fuels contain some sulfur because the living things they derived from contained sulfur.

Any discussion of the sulfur cycle must include the role played by the atmosphere. Gases from volcanic vents, smoke from industrial chimneys, and exhaust from trucks and cars, all put sulfur gases into the air around us.

Sulfur Dioxide

When any of the metal ores mentioned previously, plus many others, are heated in air, the smelly, colorless gas called sulfur dioxide, SO_2, forms. For example, when the mineral sphalerite, which is zinc sulfide, ZnS, is heated, the result is zinc oxide, a white powder, plus sulfur dioxide, a gas.

Sulfur dioxide is a two-faced gas. It is both dangerous and helpful to humans. If inhaled, it can cause painful choking and even suffocation. Too much of it in the air causes the throat and breathing passages to become irritated and close up.

On the other hand, the chemical has a number of important uses. As we have seen, it is used in bleaching. It is also useful in the manufacturing of paper pulp. Sulfur dioxide is used in a hot solution that dissolves the "glue" that holds together the cellulose fibers in wood. These fibers, then, make up paper pulp.

Interestingly, sulfur dioxide is also used to produce more sulfur. Sulfur exists in petroleum and natural gas as poisonous hydrogen sulfide, H_2S. When sulfur dioxide gas is injected into natural gas and petroleum, the sulfur dioxide takes the hydrogen from hydrogen sulfide, forming water, and leaving sulfur (which does not dissolve in water):

$$2H_2S + SO_2 \rightarrow 3S + 2H_2O$$

Until recently, the sulfur obtained this way was discarded. Now, it is collected and purified for use. More sulfur is now obtained from fossil fuels than by the Frasch method.

The Astonishing Worms

Oceanographers have known for a long time that volcanoes often formed in the oceans. In fact, the world's tropical islands were formed by volcanoes.

But not until 1977 did scientists discover that deep in the ocean other vents were spewing sulfur gases from beneath the

Red tubeworms thrive around an ocean-bottom vent that spews boiling water and sulfur.

crust. Even more astonishing was the discovery that around these hydrothermal (hot water) vents a peculiar mode of life had developed. Scientists using cameras on tiny remote-controlled submarines found boiling water escaping through these underwater vents. Yet thriving around these vents was a variety of living things unlike anything found at the upper levels of the ocean or on solid land.

These living things depend on certain bacteria's chemosynthesis, the ability to convert chemicals to energy without sunlight. The boiling water escaping from the deep ocean-floor vents contains a great deal of hydrogen sulfide. Specialized bacteria convert the sulfur in the water directly into energy for life. Other creatures depend on these bacteria for their existence.

Among the most startling of the creatures discovered are long red tubeworms that thrive through the actions of the sulfur-eating bacteria living inside them. These worms don't eat the bacteria. In fact, the worms have no digestive systems. Instead, the bacteria convert the hydrogen sulfide into chemical "food" that nourishes the tubeworms.

These deep-dwelling tubeworms are red in color because of the chemical hemoglobin, which transfers hydrogen sulfide to the bacteria. This is the same chemical that in human blood transfers oxygen from the blood vessels to the body's cells. The longest tubeworms are found in the Pacific Ocean. Such shellfish as clams, mussels, and shrimp are also found around the vents.

THE MOST IMPORTANT ACID

One of the prime uses of sulfur dioxide is to make sulfuric acid. Oxygen is added to the gas to make sulfur trioxide, SO_3, which is then reacted with water to form sulfuric acid, H_2SO_4.

$$2SO_2 + O_2 \rightarrow 2SO_3$$

$$SO_3 + H_2O \rightarrow H_2SO_4$$

Sulfuric acid was probably made by a natural sequence of such reactions in ancient times. It was first deliberately produced and written about by Geber, an alchemist in the Middle Ages. Alchemy was a combination of attempts at magic, early science, religion, and philosophy. Many of the processes that form the basis of modern chemistry were discovered by alchemists. The making of sulfuric acid was perhaps the most important.

A large fertilizer factory at night. Sulfuric acid is used in the manufacturing process.

Alchemists called sulfuric acid "oil of vitriol" because they made it out of a substance they called vitriol, which was a metallic sulfate, such as copper sulfate, $CuSO_4$, or iron sulfate, $FeSO_4$. Today, nine-tenths of all sulfur mined and processed is used in making sulfuric acid. More sulfuric acid is produced by the worldwide chemical industry than any other chemical. More than 10,000 manufacturing processes involve the use of this acid. These processes range from making textiles to making fertilizers.

Copper sulfate, one of the substances known to alchemists, is used today in a spray to kill fungi that can damage crops. Seen here is a cherry orchard being sprayed.

A Strong Acid

An acid is often defined as a substance that tastes sour and that can corrode, or burn, or even dissolve other materials. **Never taste any acid; a strong acid can kill.**

Specifically, an acid is a chemical that, when dissolved in water, releases hydrogen ions, H^+. The stronger an acid is, the more hydrogen ions it releases. For example, the acid in grapefruit is a mild acid. It releases only a few hydrogen ions. Sulfuric acid, H_2SO_4, which is thick and syrupy, is very strong. It releases many ions.

Sulfuric acid is used in huge quantities in making other chemicals and in producing steel and other metals. It requires very careful handling because of the burns it can inflict. It does not burn by heat but by rapidly drawing water from the skin.

If water is poured into sulfuric acid, it reacts with great heat that makes the acid splatter. Industrial processes involving sulfuric acid must be carried out with great care. Water should never be poured directly into any acid. It can instantly boil, splashing dangerous acid at anyone nearby.

Making Sulfuric Acid

There are several different processes for making sulfuric acid. The primary process used in the United States is called the contact process. Two gases—oxygen and purified sulfur dioxide—come into contact with a catalyst, which is a chemical that speeds up a chemical reaction but is not part of the final products. A catalyst comes out unchanged, but the chemical reaction involved has created new products. The catalyst used in processing sulfuric acid is usually vanadium pentoxide, V_2O_5 (vanadium is V, element #23).

If sulfuric acid is mixed with water, the acid breaks down and forms hydrogen sulfate ions and hydronium ions. Hydronium ions are molecules of water, H_2O, that have combined with an extra hydrogen ion, making H_3O^+.

Because there is considerable waste of sulfur dioxide in the contact process, manufacturers have developed another way to make sulfuric acid, called the double contact system. The SO_2 passes over catalysts twice, yielding more acid and less waste.

The Sulfur Sponge

As noted above, sulfuric acid burns by dehydrating, or removing water from, the skin. With skin, this dehydration is harmful, but with other processes, the dehydrating effect of the acid is useful. For example, some chemicals need to have water removed from them without affecting the chemicals themselves. Sulfuric acid reacts with the water and nothing else.

Other chemicals, such as oxygen, can also be removed from a substance by using sulfuric acid. In the steel industry, for example, sulfuric acid is used to clean metal oxide from iron and steel items before they are coated with other materials.

Superphosphates

More sulfuric acid is made than any other chemical primarily because it is used in making the largest category of chemical products—fertilizers for agriculture. Phosphate rock is used as a source of phosphorus (P, element #15), an element plants need to grow properly. Crushed rock does not dissolve easily, though, so the phosphorus in it can't be taken up by plant roots.

About 150 years ago, it was discovered that if crushed phosphate rock is treated with sulfuric acid, it dissolves more easily. The resulting product, called superphosphate, is the main phosphorus fertilizer used today.

The Acid Battery

Batteries are devices that produce a flow of electrons from various chemicals. A flow of electrons, of course, is electricity. The first person to work with electricity produced by chemicals was Italian scientist Alessandro Volta in 1794. The unit of electrical measurement called the volt is named for Volta.

For 60 years following Volta's invention, batteries had to be replaced when the chemicals in them were used up in producing electricity. Then, in 1859, French physicist Gaston Plante produced the first so-called storage battery. It was one in which the fluid, called the electrolyte, could be replenished so that the battery went on working. The electrolyte used in such storage batteries is sulfuric acid.

The storage battery used to start the engine in cars is a 12-volt battery. Within the battery are six cells, or units. Each cell is made up of a positive plate, or anode, of lead dioxide (PbO_2) and a negative plate, or cathode, of spongy lead. Both are immersed in an electrolyte bath of sulfuric acid and water.

In half the cell, atoms of lead break the two hydrogen atoms in a molecule of sulfuric acid into two protons and two electrons:

$$Pb + H_2SO_4 \rightarrow PbSO_4 + 2H^+ + 2\ e^- \text{ (electrons)}$$

In the other half of the cell, the lead dioxide reacts with the two free electrons, forming lead sulfate and water:

$$PbO_2 + H_2SO_4 + 2H^+ + 2\ e^- \rightarrow PbSO_4 + 2H_2O$$

In between these two reactions, free electrons travel through wires lighting light bulbs, playing a radio, or providing electricity to start a car's engine.

An automobile storage battery has been cut in half to show the cells in which the electricity is produced. The metal plates in each unit are surrounded by sulfuric acid, which can be replenished.

Not Wanted

Sulfuric acid, though it is important to industry, is not always wanted where it appears. You may have noticed that books produced in recent years often have the words "Printed on acid-free paper" on an inside page.

For a long time, paper was manufactured using aluminum sulfate, $Al_2(SO_4)_3$. (Aluminum is Al, element #13.) But over the years, the pages of such books began to fall apart. It was eventually realized that aluminum sulfate in the paper was combining with moisture in the air to produce sulfuric acid, which ate up the paper. Now, good-quality paper is made without acid-producing chemicals.

In an experimental program to discover ways to make biofuels from plants, sulfuric acid is used to make the tough fibers in plant matter break down more quickly when treated with steam.

A VISITOR IN THE ATMOSPHERE

A diagram of the gases that make up the blanket of air we call the atmosphere rarely includes sulfur dioxide, SO_2, and sulfur trioxide, SO_3 (together, they are referred to as SO_x). Unlike nitrogen (N, element #7), oxygen, and some other gases, the percentages of SO_x in the atmosphere are not the same worldwide.

Sulfur gets into the atmosphere from many different sources, but it rarely spreads out very far from those sources. Most of the molecules settle back to earth or decompose quickly. Sulfur dioxide molecules in the atmosphere last only about a week.

There has always been some sulfur in the atmosphere, from the earliest days of Earth when erupting volcanoes

Natural haze in the Great Smoky Mountains has a colored tinge because of sulfur and other pollutants in the air.

pockmarked the crust. That sulfur played a role in the evolution on life, so sulfur is one of the essential elements in living things.

Natural Sources

Sulfur still enters the atmosphere from natural sources. The "Ring of Fire"—the chain of volcanoes surrounding the Pacific Ocean—reveals the places where tectonic plates meet. These volcanoes spew both hydrogen sulfide and sulfur dioxide into the air, but only occasionally do they explode with enough force for the sulfur molecules to reach the upper atmosphere where global winds will carry them around the planet.

Sulfur gases also form over low-lying, non-volcanic land. When vegetation decays and is decomposed by bacteria in marshes, swamps, and bogs, where no oxygen is available, hydrogen sulfide gas is produced.

Certain marine algae emit sulfur into the atmosphere as a result of biological processes.

Another source of sulfur dioxide in the air is one that people rarely think of—ocean spray. Early in the life of our planet, the ocean was filled with acids created by sulfur, but these acids have long since turned to sulfates in the rocks beneath the sea. Today, sulfates are still the fourth most abundant constituent of seawater, after chlorine, sodium, and magnesium (Mg, element #12). As the surging ocean moves, spray is tossed into the air, where many of the elements remain in the atmosphere for a time.

Certain types of ocean-living primitive plants called algae take sulfur from the water, digest it, and give off dimethyl sulfide, $(CH_3)_2S$. Because of this, the concentration of sulfur in the atmosphere over the ocean is considerably higher than over land.

Our planet and its living things are geared to the amount of sulfur put into the atmosphere through such natural sources. But environmental problems began when humans added huge quantities of sulfur to the atmosphere through their activities.

SO_2 and the Making of Ozone

Almost all atmospheric sulfur dioxide comes from the burning of petroleum, coal, and other fossil fuels. Ever since the beginning of the Industrial Age, humans have been burning coal as a source of heat and energy. Coal was burned to boil water for steam that would turn turbines to make electricity. Factories burned coal to make heat for industrial processes. Both power plants and factories burned whatever kind of coal they could obtain cheaply. Unfortunately, much of this coal contained considerable sulfur. When this sulfur-heavy coal is burned, it produces sulfur dioxide gas, which rises into the air with smoke.

Sulfur dioxide absorbs solar radiation, which knocks an oxygen atom out of the sulfur dioxide molecule, leaving sulfur oxide and oxygen. The sulfur oxide recombines with diatomic oxygen, O_2, in the atmosphere, making sulfur dioxide again, and also releasing another free oxygen atom. These free oxygen atoms combine with O_2 to make a triatomic (three-atom) molecule, O_3, which is called ozone.

$$SO_2 + radiation \rightarrow SO + O$$

$$SO + O_2 \rightarrow SO_2 + O$$

$$O + O_2 \rightarrow O_3 \ (ozone)$$

We safely breathe O_2. In fact, we need it to live, but O_3 causes us problems. At its mildest levels, it makes our eyes red and irritated. At its worst, it can cause someone with lung problems to become very ill and even die. When weather conditions cause pollutants to stay close to the ground, many cities issue ozone alerts, warning people with lung problems to stay indoors. Ozone also damages crops.

The U.S. Environmental Protection Agency (EPA) has set standards for the amount of sulfur dioxide that can be in the air at a specific location. It calls for no more than 0.03 parts per million (ppm) averaged over a year and 0.14 ppm measured over a 24-hour period.

The ozone that people breathe is at lower atmospheric levels (the troposphere) where it can harm our health. In the upper atmosphere (the stratosphere), solar radiation forms ozone molecules. These settle in a thin layer that blocks harmful ultraviolet rays from striking the earth. In recent years, it has been found that the protective ozone layer is getting very thin, probably as a result of other pollutants. Unfortunately, the sulfur-driven ozone in the troposphere does not rise to replenish the ozone in the stratosphere.

Acid Falling from the Sky

Sulfur dioxide in the air reacts with oxygen, O_2, to form sulfur trioxide, SO_3. Sulfur trioxide doesn't stay around long because it reacts with water vapor, H_2O, and oxygen in the air to form sulfuric acid.

$$2SO_2 + O_2 \rightarrow 2SO_3$$

$$SO_3 + H_2O \rightarrow H_2SO_4$$

The same process occurs with nitrogen, which forms nitric oxide and then nitric acid. Together, the two acids can be carried long distances by wind. At some point, however, they fall to the ground in rain or snow. This is called acid precipitation.

Entire forests, often located many miles from the source of the acids, have been damaged by acid precipitation. Some lakes have

These trees have been killed by acid precipitation in the form of fog that often enshrouds the forest.

had all the life in them—both plants and animals—disappear because so much acid-containing precipitation has run into their waters. Acid precipitation also eats away limestone and erodes marble buildings and statues.

The Problem with Coal

The quickest solution to cutting down SO_2 in the atmosphere was to stop burning coal that contained a lot of sulfur. High-sulfur coal contains more than 3 percent sulfur. Low-sulfur coal contains 1 percent or less. The sulfur content in coal may be inorganic, usually in the form of pyrite, or organic, such as from the remains of the plants and animals that made up the coal.

The Clean Air Act of 1990 required coal-burning plants to use only coal with a low-sulfur content. This move did serious harm to some states' economies. Illinois, for example, is a major source of high-sulfur coal. Before the Clean Air Act, almost 11,000 miners produced more than 60 million tons of coal from that state's mines. By 1998, fewer than 4,500 miners were producing coal. More than half the state's 43 coal mines had closed.

The future may be improving, however. Scientists at power plants have been running studies on coal use. They have

Low-sulfur coal is obtained in Wyoming by strip-mining. Midwestern coal contains more sulfur than western coal.

discovered that they can mix some of Illinois's cheaper coal with low-sulfur coal (generally from western states) and still meet federal antipollution standards.

One of the most common solutions to sulfur air pollution is a device called an electrostatic precipita-tor. This device sends an electric charge into smoke before it comes out of the chimney of a power plant where coal is being burned. The electrical charge makes the sulfur particles clump together and fall out (precipitate) from the smoke. The solid par-ticles are then collected and disposed of safely.

A biological process for cleaning up coal involves sulfur-eat-ing bacteria. The bacteria that digest sulfur take about six weeks to remove about 40 percent of the sulfur from coal. Genetic engi-neers are hoping to genetically change these bacteria to speed up the process and make it more effective.

On the Road

The part of petroleum that becomes gasoline for cars contains little sulfur. In recent years, the average sulfur content of gaso-line has been about 340 ppm. In California, the law requires that gasoline contain no more than 30 ppm. In 1999, the EPA proposed that the sulfur remaining in gasoline all over the United States be lowered by 90 percent by the year 2004.

But this takes care of only part of the problem. Fuel used for diesel engines comes from a different part of petroleum. It is heavier than gasoline. It has been allowed to contain much more sulfur than gasoline—as much as 500 ppm. To make pollution problems worse, the number of diesel-engine trucks on the road has increased greatly.

Catalytic converters are devices that have long been used on cars to trap some of the nitrogen compounds in engine exhaust. A catalytic converter has a surface—the catalyst—on which harmful nitrogen oxide and carbon monoxide molecules are changed into chemicals that are harmless in the atmosphere. But sulfur also settles on the catalytic surface, taking up space needed to change the other molecules. Catalytic converters have not been used on diesel engines because the sulfur content of the fuel has been so high that they would quickly clog up the catalytic surface. So while the atmosphere has been getting cleaner because of the devices on automobile engines, diesel engines on other vehicles have continued to pollute the air.

In 2000, the Environmental Protection Agency proposed rules that would reduce the sulfur in diesel fuel by a full 97 percent. If it can be accomplished efficiently, more diesel-powered vans, pickups, and sports utility vehicles might take to the road.

Worldwide, thousands of buses and trucks, like this diesel bus in Thailand, continue to pollute the air people breathe.

This beautiful lake in the Adirondack Mountains of upstate New York has no life in it today because of acid precipitation.

Global Warming

The main environmental issue facing people of the twenty-first century is global warming. For decades, some scientists have been convinced that the base temperature of Earth is rising because of the use of so much fossil fuel by humans. Other scientists refuse to accept the idea. They insist that any warming is simply part of a natural cycle. But as the new millennium began and Earth's weather appeared to be changing, increased attention is being paid to the possibility of global warming.

Predictions about the consequences of global warming vary, but they tend to include such catastrophes as the melting of the polar ice caps and the general warming of the oceans. Warmer water takes up more space than cooler water does. Together, these changes would cause coastal cities around the world to flood. There would be less snow, providing less freshwater for drinking. The kinds of crops grown in any specific area would change. Droughts would occur in major agricultural areas.

The rising temperature of the planet appears to be the result of pollutants put into the air by burning fossil fuels. All burning puts carbon dioxide, CO_2, into the air. Burning fossil fuels also puts nitrogen oxides (NO_x) into the air. These chemicals react with normal diatomic oxygen to produce ozone. Together, SO_x, NO_x, O_3, along with methane, CH_4, and other chemicals are called greenhouse gases because these multi-atom molecules hold heat close to the planet's surface.

It's Not All Bad

Some scientists believe that we have made global warming worse by reducing the amount of sulfur in the atmosphere. Sulfur dioxide in the atmosphere reacts with water vapor, H_2O, to form tiny particles called aerosols. Aerosols reflect sunlight back into space, cooling the area underneath. By cutting down on sulfur dioxide in the air, we have prevented aerosols from forming.

Sulfur in the atmosphere also has other values. It falls to the ground in rain and snow, returning sulfur to the soil. Farmers rarely have to concern themselves with adding sulfur fertilizer to their fields. However, in recent years, much less rain and snow have fallen than usual in many areas, and many agricultural regions are suffering from drought. When this happens, farmers have to add sulfur to the soil for their crops, especially corn. Those farmers who plow corn stalks and other crop waste into the ground instead of clearing it away do not have to add sulfur because the decomposed plants return sulfur to the soil.

Global Sulfur and Dinosaurs

When volcanic eruptions occur with great force, sulfur is sent high into the upper atmosphere. The resulting aerosols can affect the climate on the entire planet. When the volcano Mount Pinatubo erupted in the Philippines in 1991, it sent sulfur and other debris 30 kilometers (18 miles) into the atmosphere. The aerosols that formed circled the globe on high-altitude winds. For more than two years, these aerosols affected Earth's weather.

Some paleontologists (scientists who study fossils) have suggested that this same process was an important factor in

Many scientists think that dinosaurs became extinct about 65 million years ago because a meteor struck Earth, vaporizing the sulfur in the ground. It entered the atmosphere and became sulfuric acid, which destroyed the green plants dinosaurs needed to survive.

causing the extinction of the dinosaurs. They now believe that a meteor struck Earth on the Yucatan peninsula of Mexico about 65 million years ago, causing changes that revolutionized life on Earth.

One of the changes that scientists think might have happened was brought about by the sulfur that was in the ground where the meteor struck. That sulfur was vaporized by the heat of the meteor. The sulfur then entered the atmosphere, linked up with water vapor, and formed sulfuric acid. The sulfuric acid fell to the surface as acid rain, destroying the green living plants on which dinosaurs depended. Other animals that did not depend directly on living plants survived. With the dinosaurs gone, mammals had a chance to evolve.

SKUNKS, GARLIC, AND PROTEINS

Perhaps you have been riding in a car and suddenly wrinkled your nose at the awful odor of skunk in the air. Skunks, small black-and-white members of the weasel family, have two glands under the tail. When threatened, they often first make agitated movements like stamping their feet. If the threat does not go away, they turn tail to the enemy—not to run, but to voluntarily spray a terrible-smelling liquid at the enemy. The spray can travel 3 meters (10 feet) or more, driving the enemy away. The scent can linger in the air for several days afterward.

The skunk's offensive odor comes from sulfur compounds—not just one, but three. One is the same dimethyl sulfide given off by algae. The other two are chemicals called mercaptans or thiols.

A skunk raises its tail to spray an enemy with awful-smelling sulfurous liquid produced in glands under its tail.

All Living Things

The skunk and its sulfurous stink are among the more noticeable examples of sulfur in living things, but actually the element occurs in all plants and animals. Sulfur is one of the essential elements required by all living things. It is a *macronutrient*, meaning that it must be obtained in foods in fairly large amounts.

Most living things get their energy in the process of oxidation, combining molecules from food with oxygen—in other words, "burning" the food. Scientists have suggested that primitive plants used sulfur rather than oxygen for this same process. Though most living things have replaced sulfur with oxygen, some species still depend on sulfur, such as the bacteria that "feed" the tubeworms in the bottom of the ocean.

Bacteria and the Power of Purple

Tubeworms and swamps have the so-called sulfur bacteria in common. These bacteria remove oxygen from sulfur compounds and then give off hydrogen sulfide.

There are two different kinds of sulfur bacteria. One kind changes sulfides into sulfates by taking oxygen from their environment. These bacteria are useful in soil, where they turn sulfur compounds into the sulfates that plants can utilize. They are called aerobic bacteria, meaning that they require oxygen to live.

The other kind of sulfur bacteria live where there is little or no oxygen. Called anaerobic bacteria, they break down sulfur compounds and give off hydrogen sulfide, H_2S. The decomposition of plant matter in water-logged ground, such as in marshes, is the work of such bacteria.

The differences between the two kinds of sulfur bacteria can deliberately be put to work. The waste products of cattle and pigs are often dumped into waste ponds, where the owners count on aerobic bacteria gradually to decompose the manure into a less smelly form. However, anaerobic bacteria in the waste

This animal-waste pond on a farm looks purple because it contains special purple sulfur-eating bacteria.

pond produce very smelly hydrogen sulfide. The terrible smell given off often causes complaints from neighbors.

Scientists are now adding another bacteria called purple sulfur bacteria to waste ponds. These bacteria have the ability to take in hydrogen sulfide and store the resulting molecules of elemental sulfur within cells until the sulfur is oxidized. Waste lagoons where these bacteria thrive actually turn purple from the purple sulfur bacteria.

Builders of Protein

Carbon is the prime element of life as we know it because carbon atoms can make more compounds than any other elements. The atoms of most other elements can attach themselves to only a few identical atoms, perhaps forming a ring or a three-dimensional triangle. But carbon atoms can attach themselves to one another in long chains, with other elements attaching themselves at the sides of the chain in many thousands of combinations. This variety of carbon-containing, or organic, compounds is what makes possible all the many different molecules that make up living things.

Sulfur is one of the few elements that behaves like carbon in making long chains. It fits in nicely with carbon in living things.

Living things are made up mainly of huge, complicated molecules called proteins. They, in turn, are made of smaller units called amino acids. There are at least 20 amino acids that combine in many different ways to make the many different proteins. Sulfur is present in several amino acids. This sulfur is organic sulfur—its compounds contain carbon. Sulfur also exists in living things in inorganic (without carbon) form. This sulfur occurs as sulfide ions, S^{2-}.

Each amino acid contains a basic part that is the same in all amino acids. But they differ in the atoms that are in one section called a side chain. The amino acids also differ in how the atoms in that side chain are arranged.

Several common amino acids include sulfur in the side chain. These sulfur-containing amino acids are important molecules in enzymes, which are the substances that act as catalysts for the vast number of chemical processes that take place in the body.

A disulfide (two-sulfur) bridge often connects amino acids in proteins. This is a bond between two cysteines.

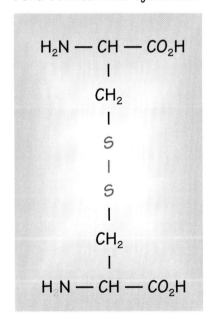

The main sulfur-containing amino acid is methionine. We must get methionine in food. It is found in many foods, including eggs, meat, milk, and various cereals. The white of an egg contains a great deal of methionine. Methionine helps produce another sulfur-containing amino acid, cysteine. Cysteine is found in many proteins that protect the body, such as skin, hair, nails, and horns.

Sometimes two cysteine molecules are connected to each other by two sulfur atoms. This connection is called a disulfide bridge. One of the most important disulfide bridges in the human body is in insulin, the hormone made in the pancreas that regulates the body's use of sugar. A person whose pancreas fails to make enough insulin has a disease called diabetes.

Human Odors

Almost all compounds of sulfur have a foul, or at least an unpleasant, smell. And humans have their share of these compounds. For example, sebum is the oil produced by the tiny glands in our skin. When it reaches the surface of the skin, this oil breaks down into various substances. Some of the compounds in sebum contain sulfur, causing an odor.

Methyl mercaptan, one of the ingredients in skunk odor, is the base molecule of bad breath, or halitosis. Methyl mercaptan has been called one of the worse smells in the entire chemical world. It is formed in the body when bacteria break down natural proteins and then decompose two of the sulfur-containing amino acids, cysteine and methionine. The final result is methyl mercaptan, which enters the bloodstream and is carried to the lungs, where the foul-smelling chemical is exhaled. Some chemicals that break up sulfur compounds are available for human consumption, thus eliminating bad breath.

When methyl mercaptan is formed by dead skin decomposing in the closed environment of tight shoes, foot odor results. Lastly, mercaptans are given off by dead animals rotting. This is one of the most awful—and memorable—of all odors.

Plant Life

Sulfur plays an important role in chlorophyll, the green matter in plants that carries on photosynthesis. When there is not enough sulfur for plants to use, they gradually turn lighter green in color, which means they are losing their efficiency at producing their own food.

The lighter areas between veins show that this plant is suffering from a lack of sulfur.

Sulfur is added to soil to make it acid. As we saw previously, an acid is a chemical that, when added to water, releases hydrogen ions, H^+. Some plants, especially rhododendrons and azaleas, like acid soils. They benefit from the hydrogen ions released from the soil when rain washes through it. Most plants, though, do not like acid soil. If a garden or farm field is near an industrial plant that produces acid smoke, the soil may turn acidic. Gardeners add lime, $Ca(OH)_2$, to the soil to make the soil nonacidic, or neutral.

To be used by plants, sulfur must first be digested by bacteria. This process oxidizes the sulfur, changing it to sulfate ions, SO_4^{2-}. Sulfate is a form of sulfur a plant can absorb. However, in the process of being converted by bacteria, hydrogen ions are released from water, making the soil acidic.

Garlic, Onions, and Eggs

Sulfur in the foods we eat provides both flavor and odors (good and bad). Garlic and onions, which many people regard as indispensable in many recipes, get their recognizable flavor and odor from several complex volatile sulfur compounds (VSC). *Volatile* means "easily evaporated."

Garlic has been regarded by many people for thousands of years as being important to health. For a long time, the benefits of this relative of the onion were thought to be imaginary. Recently, though, researchers have studied the plant's chemistry and concluded that garlic can play a role in preventing blood clots, digestive problems, and heart disease.

The penetrating odor of garlic comes from its sulfur compounds, especially mercaptans. The garlic bulb itself does not smell until it is cut. Chemists have found as many as nine different compounds that are released when a garlic bulb is cut. The compounds emitting the odor of garlic do not survive cooking.

When raw garlic is eaten, the complex sulfur compounds are

broken down into simpler ones. These travel through the blood-stream and are exhaled through the lungs, giving a person "garlic breath." It takes only a little garlic to cause "garlic breath." A person standing near someone who has eaten garlic can detect as little as 1 part per billion (ppb) in the garlic eater's breath.

Some other vegetables also have "bad" odors. The vegetables often called "kale vegetables" include broccoli, cabbage, brussel sprouts, and cauliflower. They are notorious for the odors they give off as they are cooked. These odors also result from volatile sulfur compounds.

Since hydrogen sulfide is often described as having the smell of rotten eggs, clearly eggs

These garlic plants have just been harvested. The flavor—and the odor—is in the bulb. The odor, though, is not released until the bulbs are cut.

contain considerable sulfur. An egg boiled too long may form a greenish layer on the outside of the yolk. This color is caused by sulfur.

We need a regular supply of sulfur in our diets because it is an essential ingredient in the proteins our bodies use to grow, maintain, and repair cells. Three B-complex vitamins contain sulfur—thiamine, biotin, and pantothenate.

Fortunately, most of what we eat, as long as it is strong in proteins, provides the sulfur we need. However, it is useful to know what foods are rich in sulfur. Meats, dairy products, nuts, grains, and legumes (beans) are the best sources of sulfur.

DRUGS
AND
DETERGENTS

You are a big user of the element sulfur, and you probably did not know it. So is every man, woman, and child in the United States. That is because sulfur is used in the most important products of the chemical industry. It has been estimated that more than 100 pounds (45 kilograms) of sulfur are used each year for every individual in the nation.

Most sulfur is used to make sulfuric acid for the production of fertilizer. The rest, though, has many uses that touch our lives every day.

Good for What Ails You

Water as it comes from a well or a lake is not just H_2O. It actually contains many other chemicals, especially those that are dissolved by rainwater from rocks and soil and carried into the water. Most of these dissolved minerals are ions.

The ancient Romans built this once beautiful and elaborate spa in Bath, England, to take advantage of a mineral-filled spring.

The dissolved sulfur in water is an anion (negative ion), SO_4^{2-}. Other anions include chloride, Cl^-, and bicarbonate, HCO_3^-. Magnesium sulfate, $MgSO_4$, and sodium sulfate, Na_2SO_4, are the main sulfates in freshwater.

Since ancient times, some people have been convinced that drinking—or at least bathing in—the waters of hot springs is good for them. Health resorts called spas were built around such mineral-filled hot springs. The ancient Romans who went to Britain founded the town of Bath around a hot mineral spring. The water of such spas usually smells and tastes of sulfur. White Sulphur Springs, West Virginia, grew up around such a spring.

Fighting Disease

In the nineteenth century, mothers who wanted their children to fight off colds and have lots of energy often gave their children a spoonful of a combination of sulfur and molasses. The molasses would hide the taste of the sulfur. Whether this concoction really did any good is uncertain, but the mothers usually had confidence in it.

Better medicinal uses for sulfur were discovered in the twentieth century. The German physician and biochemist Gerhard Domagk is credited with starting the era of great miracle drugs in 1932 with his discovery of sulfa drugs.

Domagk was making a survey of the uses of dyes when he saw that one of them, a red dye with the trade name Prontosil, contained sulfur and seemed to slow infections caused by the *Streptococcus* bacteria in rats. Domagk experimented on a human when his own daughter became seriously ill from a strep infection. His "sulfa" drug cured her. Although sulfa drugs are not strictly antibiotics, they are often regarded as such. Domagk received the 1939 Nobel Prize for Medicine for his discovery.

Soon after Domagk's discovery, Daniele Bovet, a scientist working at the famed Pasteur Institute in Paris, found that only

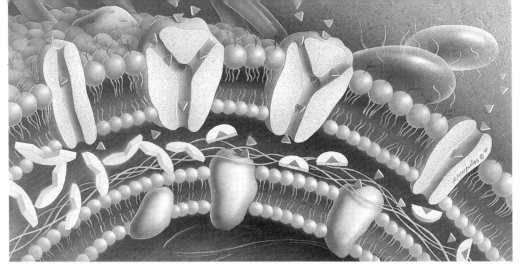

Sulfa drugs keep bacteria from reproducing by weakening the membrane around the nucleus of the bacterial cell (shown here as the bottom curve).

part of the Prontosil chemical was actually involved in curing strep infections. It was the part called sulfanilamide. This discovery led to the development of a series of drugs called sulfonamides, or sulfa drugs.

Sulfa drugs saved many lives in World War II and in the years before penicillin and other antibiotics became easily accessible. However, about 3 out of every 100 people are allergic to sulfonamides. Interestingly, they are usually not allergic to other drugs that contain sulfur, such as the antibiotic amoxicillin.

Personal Care

A sulfate called lauryl sulfate is a major ingredient in many personal-care items, especially common, everyday soap. Lauryl sulfates are surfactants (short for "surface-active agents"), meaning that their chemical structure breaks the surface tension that holds water and dirt-containing oils together. When surface tension is broken, the dirt is released and can be washed away. Lauryl sulfates are also used in shampoos and toothpaste.

Sulfur has long been used in fighting dandruff, the scalp condition that makes white flakes of dead skin collect on the hair. Sulfur is often combined with salicylic acid (the prime ingredient in aspirin) to treat not only dandruff but also acne.

Sulfur is also an ingredient used in products involved in the process of permanently waving hair. Both skin and hair are made up of a protein called keratin. Chains of proteins form the keratin, and these proteins are held together by two sulfur atoms in a disulfide bond. Such a bond makes skin and hair very tough.

The chemical used in giving hair a permanent wave breaks the disulfide bonds in keratin so that the protein strands can move past each other into a new position, which is determined by rollers put in the hair. Another chemical—hydrogen peroxide, H_2O_2—restores the disulfide bonds to different pairs of atoms. The hair stays curly until it grows out.

Making Soaps Work

Water with a lot of dissolved minerals in it, especially calcium and magnesium, is called hard water. Soap does not work well in hard water, so modern detergents, which do work in hard water, were invented.

Among the earliest synthetic detergents made were ones called ABS, short for alkyl-benzene-sulfonates. ABS detergents were popular because they could make oily dirt on clothing break apart and float away. But they had a major problem—they did not decompose, or degrade, after use. The ABS molecules held together through sewage-treatment processing and were still lathering up when treated water was released into lakes and

Permanent-wave chemicals break and shift the disulfide bond in keratin.

Natural Hair

Wave Lotion Containing HSCH$_2$COOH

Setting Lotion Containing H$_2$O$_2$

Waved Hair

rivers. Suddenly these bodies of water were foaming!

The manufacturers hurried to develop biodegradable detergents that would be broken down by microbes in the environment. These new detergents did break down after they were used, but one of the end products was the sulfate ion, SO_4^{2-}. Some scientists think sulfate buildup is also a problem, though this has not been proved.

Sulfur in the Food Industry

Sulfur dioxide, though it is thought to be dangerous in the environment, is used more safely in food processing. It seems to be harmless when swallowed. It is often used to preserve dried fruits. It also prevents some other foods, such as jellies and dehydrated potatoes, from turning brown in their container.

Meat is an important source of the vitamin thiamine. Because sulfur dioxide destroys thiamine, it is not generally allowed in the processing of meats. However, it is sometimes used in such meat products as sausages, where its ability to preserve the food is more important than the thiamine content of the sausage.

For some years, restaurants treated vegetables on their salad bars and potatoes used for french fries with various chemicals called sulfites. The term "sulfite" actually covers several different forms of sulfur oxide, especially sodium sulfite, Na_2SO_3. The purpose was to prevent lettuce and other foods from turning color in the air and thus not looking fresh. But this practice harmed some people.

Some people are missing an enzyme called sulfite oxidase, which converts sulfites to harmless sulfates. An estimated 1 percent of the population does not have the enzyme. For them, exposure to sulfites, as in foods, can cause fatigue, headaches, itching, and more serious allergic reactions. A consumer group brought the public's attention to the possible danger of adding sulfites to certain foods, and use of the chemicals was stopped.

Riding on Sulfur

Rubber was originally found in South America, where it was made from the sap of the rubber plant. It first became known to Europeans when French geographer Charles-Marie de La Condamine sent home samples from the Amazon jungle in the 1730s. Perhaps 75 years later, English chemist Joseph Priestley gave the substance the name *rubber* after discovering that it could rub out pencil marks.

Rubber remained primarily a children's toy and a coating that waterproofed cloth until American inventor Charles Goodyear started to tinker with it. Goodyear was looking for a way to correct rubber's bad trait of hardening and cracking in cold weather and becoming sticky in hot weather. He found that if he added sulfur to raw rubber and heated the mixture, the result was a material that was strong and flexible under common temperature conditions. He called the process vulcanization after Vulcan, the Roman god of fire.

The addition of sulfur to rubber is not just the physical process of mixing in sulfur. Chemically, rubber consists of long chains of molecules that slide on each other, making rubber rubbery. When sulfur is added to rubber, the molecules change so that they no longer slide. The substance hardens.

A painting of young Charles Goodyear experimenting with rubber on the stove in his laboratory. He invented the process of vulcanization in 1839.

A SULFUR CATALOG

The Mystery Planet

Venus, the second planet from the sun in our solar system, was known to ancient humans as just a bright light in the sky. Even when people looked through early telescopes at Venus, nothing was seen but a dark mass of swirling clouds that reflected sunlight, making the planet visible. Eventually, those clouds were found to consist mainly of sulfuric acid, H_2SO_4.

Little more was known about Venus until the Space Age, when the Soviet Union's Venera spacecraft descended through the clouds and transmitted information and photographs of the planet's surface. It turned out that the clouds of sulfuric acid encased another, quite different atmosphere close to the surface. That atmosphere is 96 percent carbon dioxide. Venus's thick clouds trap solar heat even more effectively than Earth's atmosphere does. Venus has a surface temperature greater than 460°C (860°F).

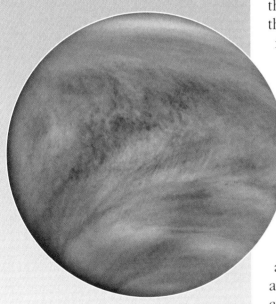

The swirling clouds around the planet Venus are made of sulfuric acid. Beneath the acid is an atmosphere of carbon dioxide.

The atmospheric pressure on the planet's surface is 90 times greater than the atmospheric pressure on Earth. Most of the relatively flat areas on the surface are lava fields from past volcanoes.

Sulfurous Satellites

The planet Jupiter was studied by the Italian astronomer Galileo in 1609 when he turned his newly invented telescope on the giant planet and discovered that it has four moons (many others were discovered later). The one he called Io is 3,218 kilometers (2,000 miles) in diameter, about the size of Earth's moon.

In the late twentieth century, an American spacecraft named *Galileo,* after the Italian astronomer, studied Jupiter and Io again. It revealed that Io is one of the few satellites in our solar system known to have active volcanoes—at least 35 of them. Their lava flows have given this moon a variety of yellow, red, and orange colors.

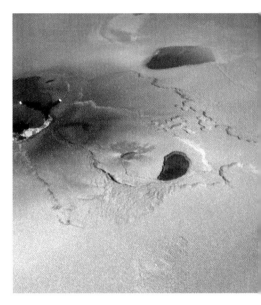

Volcanoes on Io, a satellite of Jupiter, have made the surface yellow with sulfur.

Sulfur on the surface of Io has reacted with the small amount of oxygen in the satellite's atmosphere to create a suffocating atmosphere of poisonous SO_2.

Jupiter itself has some hydrogen sulfide, H_2S, in its spectacular atmosphere. Other moons of Jupiter are known to have frozen sulfuric acid on their surfaces.

Blue Stones

During the Middle Ages, a semiprecious vivid-blue stone called lapis lazuli was used to make a brilliant blue paint. This paint was used in some of the world's great paintings and in

Lapis lazuli is a gemstone formerly used to make ultramarine pigment.

illustrating manuscripts. Lapis is not actually a mineral in itself, as most precious and semi-precious stones are. Instead, it is limestone that is colored by the mineral lazurite. The more sulfides in lapis, the deeper the blue color is.

Turning lapis lazuli into the usable pigment called ultramarine blue was a long process, requiring several weeks and at least 49 steps. This long work made ultramarine the most expensive pigment that an artist could buy. Through the centuries, synthetic versions of ultramarine were made and sold, but they never gave the beautiful deep blue made from lapis lazuli.

Driving on Sulfur

In addition to driving cars with vulcanized, sulfur-containing tires, we also sometimes drive on sulfur in the roads beneath the tires. Many roads and parking lots are paved with asphalt, which is a thick, black, sticky fluid that fills in cracks when spread hot onto a road surface. Originally a material found in the ground, asphalt is now made from the residue of petroleum refining.

Asphalt is not very long-wearing. It usually has to be reapplied to roads every few years, perhaps even annually on busy roads. However, engineers have discovered that if sulfur is added to the asphalt, it lasts longer. Called sulfur-extended asphalt (SEA), the material has the added benefit of using less petroleum and more sulfur. We are running out of petroleum, but we have lots of sulfur.

Fueling Up with Algae

Because we are running out of petroleum, researchers at the National Renewable Energy Laboratory in Golden, Colorado, and the University of California at Berkeley, California, have been looking at different ways to save our petroleum resources. In one stunning discovery, they have found that they can make some algae produce hydrogen instead of the oxygen that green plants usually produce. The gas these plants give off can be collected and used to power engines.

First, the scientists feed the algae sulfur, which the primitive, one-celled plants incorporate into all their proteins and other molecules during photosynthesis. The usual final product of photosynthesis is oxygen. However, if the researchers cut off the supply of sulfur for 24 hours, an enzyme develops that makes the algae produce hydrogen instead of oxygen. The hydrogen gas can then be collected.

A little bit of hydrogen produced in a laboratory is a long way from plants producing enough hydrogen to run a vehicle, but these scientists think that hydrogen-making algae ponds may be in the future backyards of many drivers.

Scientists have found that feeding algae with sulfur and then cutting off the supply can make the algae produce hydrogen instead of oxygen.

Sulfur in Brief

Name: Sulfur
Symbol: S
Discoverer: Known since ancient times
Atomic number: 16
Atomic weight: 32.06
Electrons in shells: 2, 8, 6
Group: 16 (sometimes called VIA); other elements in Group 16 include oxygen, selenium, tellurium, and polonium
Usual characteristics: Depend on the form, or allotrope. The most frequent form is dark-yellow rhombic crystals. If heated, they change into lighter yellow monoclinic crystals. At least 17 other allotropes exist, including amorphous, or plastic, sulfur
Density (mass per unit volume) of rhombic: 2.07 g/cm³
Melting point (freezing point) of rhombic: 112.8°C (235°F)
Boiling point (liquefaction point) of rhombic: 444.6°C (831°F). If sulfur is cooled quickly from above the boiling point, it sublimes, or changes directly into a solid.
Abundance:
    Universe: 9th most abundant element; about equal to iron
    Earth: 5th most abundant element
    Earth's crust: 0.03%; 12th most abundant element over all, but third most abundant element in minerals
    Earth's atmosphere: Minimal, varies depending on location and human activity
    Human body: 0.25%
Stable isotopes (sulfur atoms with different numbers of neutrons in their nuclei): 95% of all natural sulfur is S-32. Remainder is S-31, 33, and 36
Radioactive isotopes: S-30, 31, 35, 37, 38, 39, and 40

Glossary

acid: definitions vary, but basically it is a corrosive substance that gives up a positive hydrogen ion (H+), equal to a proton when dissolved in water; indicates less than 7 on the pH scale because of its large number of hydrogen ions

alchemy: the combination of science, religion, and magic that preceded chemistry

alkali: a substance, such as an hydroxide or carbonate of an alkali metal, that when dissolved in water causes an increase in the hydroxide ion (OH-) concentration, forming a basic solution

allotrope: an alternative structure of an element when it exists in two or more forms. Sulfur has at least 19 allotropes.

anion: an ion with a negative charge

atom: the smallest amount of an element that exhibits the properties of the element, consisting of protons, electrons, and (usually) neutrons

base: a substance that accepts a hydrogen ion (H+) when dissolved in water; indicates higher than 7 on the pH scale because of its small number of hydrogen ions

boiling point: the temperature at which a liquid is converted into a gas, or a solid changes directly (sublimes) into a gas; also, the temperature at which a gas or vapor condenses into a liquid or solid

bond: the attractive force linking atoms together in a molecule or crystal

catalyst: a substance that causes or speeds a chemical reaction without itself being consumed in the reaction

cation: an ion with a positive charge

chemical reaction: a transformation or change in a substance involving the electrons of the chemical elements making up the substance

compound: a substance formed by two or more chemical elements bound together by chemical means

crystal: a solid substance in which the atoms are arranged in three-dimensional patterns that create smooth outer surfaces, or faces

decompose: to break down a substance into its components

density: the amount of material in a given volume, or space; mass per unit volume; often stated as grams per cubic centimeter (g/cm³)

diatomic: made up of two atoms

dissolve: of two substances, to mix to form a solution

distillation: the process in which a liquid is heated until it evaporates and the gas is collected and condensed back into a liquid in another container; often used to separate mixtures into their different components

disulfide bond: a link between two sulfur atoms, important in the structure of many proteins

electrode: a device such as a metal plate that conducts electrons into or out of a solution or battery

electrolysis: the decomposition of a substance by electricity

electrolyte: a substance that conducts electricity when dissolved in water or when liquefied

element: a substance that cannot be split chemically into simpler substances that maintain the same characteristics. Each of the 103 naturally occurring chemical elements is made up of atoms of the same kind.

enzyme: one of the many complex proteins that act as biological catalysts in the body

evaporate: to change from a liquid to a gas

fossil fuel: petroleum, natural gas, or coal, all of which were formed from the remains of plants and animals

gas: a state of matter in which the atoms or molecules move freely, matching the shape and volume of the container holding it

group: a vertical column in the Periodic Table, with each element having similar physical and chemical characteristics; also called chemical family

hormone: any of various secretions of the endocrine glands that control different functions of the body, especially at the cellular level

inorganic: not containing carbon

ion: an atom or molecule that has acquired an electric charge by gaining or losing one or more electrons

isotope: an atom with a different number of neutrons in its nucleus from other atoms of the same element

mass number: the total of protons and neutrons in the nucleus of an atom

melting point: the temperature at which a solid becomes a liquid

metal: a chemical element that conducts electricity, usually shines, or reflects light, is dense, and can be shaped. About three-quarters of the naturally occurring elements are metals.

metalloid: a chemical element that has some characteristics of a metal and some of a nonmetal; includes some elements in Groups 13 through 17 in the Periodic Table

molecule: the smallest amount of a substance that has the characteristics of the substance and usually consists of two or more atoms

neutral: 1) having neither acidic nor basic properties; 2) having no electrical charge

neutron: a subatomic particle within the nucleus of all atoms except hydrogen; has no electric charge

nonmetal: a chemical element that does not conduct electricity, is not dense, and is too brittle to be worked. Nonmetals easily form ions, and they include some elements in Groups 14 through 17 and all of Group 18 in the Periodic Table.

nucleus: 1) the central part of an atom, which has a positive electrical charge from its one or more protons; the nuclei of all atoms except hydrogen also include electrically neutral neutrons; 2) the central portion of most living cells, which controls the activities of the cells and contains the genetic material

organic: containing carbon

oxidation: the loss of electrons during a chemical reaction; need not necessarily involve the element oxygen

pH: a measure of the acidity of a substance, on a scale of 0 to 14, with 7 being neutral. pH stands for "potential of hydrogen."

photosynthesis: in green plants, the process by which carbon dioxide and water, in the presence of light, are turned into sugars

pressure: the force exerted by an object divided by the area over which the force is exerted. The air at sea level exerts a pressure, called atmospheric pressure, of 14.7 pounds per square inch (1013 millibars).

protein: a complex biological chemical made by the linking of many amino acids

proton: a subatomic particle within the nucleus of all atoms; has a positive electric charge

radioactive: of an atom, spontaneously emitting high-energy particles

reduction: the gain of electrons, which occurs in conjunction with oxidation

respiration: the process of taking in oxygen and giving off carbon dioxide

salt: any compound that, with water, results from the neutralization of an acid by a base. In common usage, sodium chloride (table salt)

shell: a region surrounding the nucleus of an atom in which one or more electrons can occur. The inner shell can hold a maximum of two electrons; others may hold eight or more. If an atom's outer, or valence, shell does not hold its maximum number of electrons, the atom is subject to chemical reactions.

solid: a state of matter in which the shape of the collection of atoms or molecules does not depend on the container

solution: a mixture in which one substance is evenly distributed throughout another

sublime: to change directly from a solid to a gas without becoming a liquid first

synthetic: created in a laboratory instead of occurring naturally

ultraviolet: electromagnetic radiation which has a wavelength shorter than visible light

valence electron: an electron located in the outer shell of an atom, available to participate in chemical reactions

vitamin: any of several organic substances, usually obtainable from a balanced diet, that the human body needs for specific physiological processes to take place

For Further Information

BOOKS

Atkins, P. W. *The Periodic Kingdom: A Journey into the Land of the Chemical Elements.* NY: Basic Books, 1995

Emsley, John. *Molecules at an Exhibition: Portraits of intriguing materials in everyday life.* Oxford: Oxford U Press, 1998

Heiserman, David L. *Exploring Chemical Elements and Their Compounds,* Blue Ridge Summit, PA: Tab Books, 1992

Hoffman, Roald, and Vivian Torrence. *Chemistry Imagined: Reflections on Science.* Washington, DC: Smithsonian Institution Press, 1993

Newton, David E. *Chemical Elements: From Carbon to Krypton.* 3 volumes. Detroit: UXL, 1998

CD-ROM

Discover the Elements: The Interactive Periodic Table of the Chemical Elements, Paradigm Interactive, Greensboro, NC, 1995

INTERNET SITES

Note that useful sites on the Internet can change and even disappear. If the following site addresses do not work, use a search engine that you find useful, such as:
Yahoo:

  http://www.yahoo.com

or Google:

  http://google.com

or Encyclopaedia Britannica:

  http://britannica.com

A very thorough listing of the major characteristics, uses, and compounds of all the chemical elements can be found at a site called WebElements:

  http://www.web-elements.com

Many subjects are covered on WWW Virtual Library. It also includes a useful collection of links to other sites:

  http://www.earthsystems.org/Environment/shtml

INDEX